Bibliographic information published by the German National Library:

The German National Library lists this publication in the National Bibliography;
detailed bibliographic data are available on the Internet at http://dnb.dnb.de .

Imprint:

Copyright © 2015 GRIN Verlag, Open Publishing GmbH
Print and binding: Books on Demand GmbH, Norderstedt Germany
ISBN: 978-3-668-08103-1

This book at GRIN:

http://www.grin.com/en/e-book/308297/the-interactions-of-beta-particles-with-
matter-the-mass-attention-coefficient

Farheen Banu Dawadi et al.

The Interactions of Beta Particles with Matter. The Mass Attention Coefficient and Particles Producing Bremsstrahlung

GRIN Publishing

INTERACTIONS OF BETA PARTICLES WITH MATTER

Dedicated to my Parents ,Teachers and Friends

By: F. K. Dawadi, Bhagyashree B. H., Chandrakala V. H., Kavita M. A., Prabhavati B. V., Shruti I. B., Arunkumar, Ashwini B., Chetana M. H. and Priyanka G. H.

Experimental Interpretation of interactions of beta particles with matter

The apparatus used in this experiment is the GM counter with its operating voltage set to 540volts.The block diagram of the experimental apparatus is as shown below;

Experimental setup:

Block Diagram of the experimental arrangement:

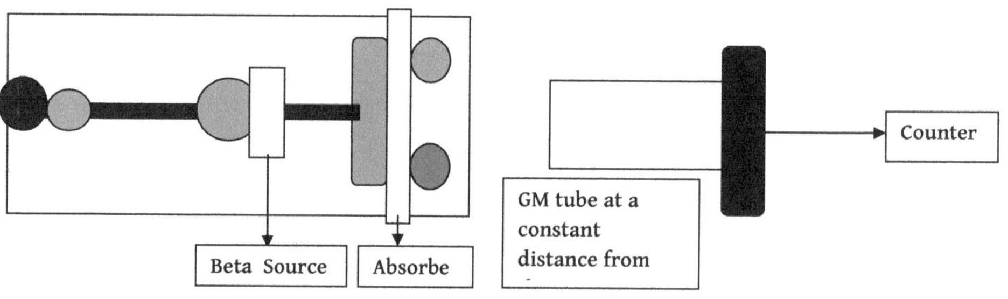

Experiment 1:

AIM: To find the mass-attenuation co-efficient of metals and compounds using beta source Tl-204:

Formula: Mass attenuation co-efficient

$$(\mu/\rho) = 1.52\,(Z^{4/3})\,(1/AE^{1.485}) \qquad cm^2\,gm^{-1}$$

Z = Atomic number

A = Mass number

E = End point energy of Thallium source

Experimental arrangement:

5.1.2: Experimental Procedure:

1. Connect the GM-tube to the digital counter/power supply and switch on.
2. Set the operating voltage to 540Volts.
3. Without the radioactive source present find the background count for 60 seconds .
4. Set up the apparatus as in the diagram(with the thinnest absorber in place)with the very fragile and expensive front of the GM tube at a known constant distance from the source.
5. Record the count received by the counter over a period of 60 seconds.
6. Repeat stages 3 to 4 with the increasing thickness of absorber.
7. Plot a graph of corrected count against absorber thickness.

8. Use graph to determine the half value thickness of the absorber (i.e.,the thickness of absorber required to reduce the number of beta particles byhalf of its counts without any absorber.This is known as the half-thickness $x_{1/2}$).

9. Also plot the graph of ln I_t V/s corrected thickness. The slope of this straight line gives the mass attenuation coefficient of the absorber used.

10. Repeat the procedure for different absorber materials.

5.1.3: Absorbers used ;

METAL NUMBER (Z)	ATOMIC	DENSITY in(ρ) gm/cm^3
1. Silver	47	10.49
2. Iron	26	7.87
3. Steel	29.16	7.8
4. Brass	29.40	8.470
5. Copper	29	8.96
6. Aluminum	13	2.7

Source used;

Thallium-204

End point energy, E= 0.764

Range of electrons;

$R = 4.07 \times (E_m)^{1.38}$

$= 4.07 \times (0.764)^{1.38}$

$= 2.81 \text{ kg m}^{-2}$

I) Attenuation of beta particle from Tl-204 using Silver as target;

Back Ground Counts for 60 seconds:

	I	II	III
	23	17	25

Mean	21.66

Density of Ag = 10.49gm/cm^3.

Thickness In cm.	Corrected Thickness in gm/cm^2	Counts for 60 seconds				Corrected counts	ln(I_t)
t	t'=ρt	I	II	III	Mean(I_0)	$I_t=I_0-I_b$	
0	0	22562	22774	2254	22625	22603.34	10.025
0.004	0.0419	19190	19040	19023	19084	19062.34	9.855
0.008	0.0839	16411	16313	16482	16402	16380.34	9.703
0.012	0.1258	13796	14002	13808	13868	13846.34	9.535
0.016	0.164	11000	11000	11001	11000	10978.34	9.303

1.Graph of I_t v/s t' ;

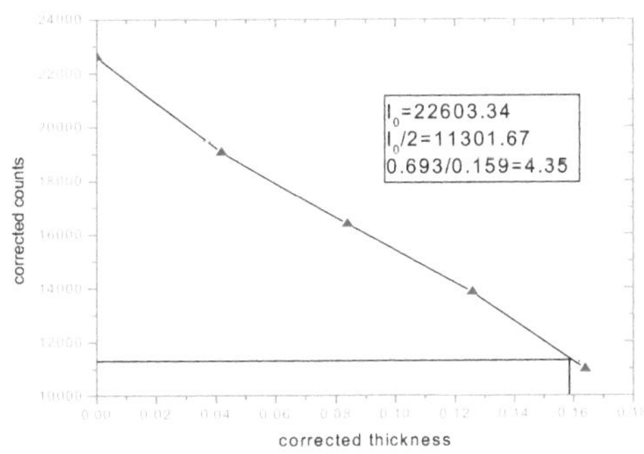

2. Graph ln(I_t) v/s corrected thickness (t');

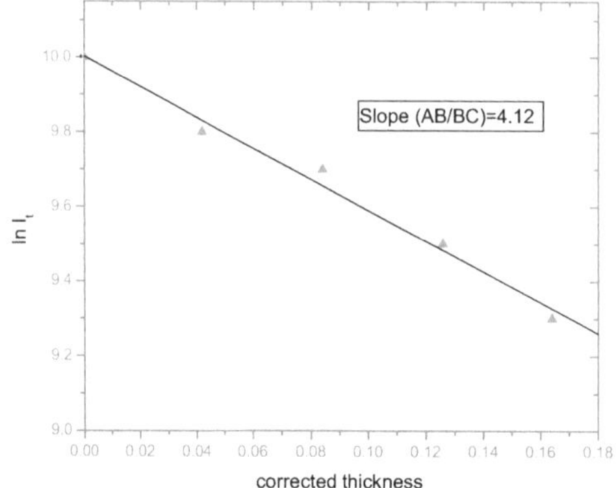

Mass Attenuation coefficient ;

μ_m =   4.35 cm^2/gm   (From graph 1)

μ_m =   4.13 cm^2/gm   (From graph 2)

Linear attenuation ;

μ = ρ x μ_m = 10.49 x 4.35 = 45.63 cm^{-1}

ii) Attenuation of beta particle from Tl-204 using Iron as target;

Back Ground Counts for 60 seconds:

I	II	III
13	19	15

Mean	15.66

Density of Fe = 7.87 gm/cm^3.

Thickness In cm.	Corrected Thickness In gm/cm^2	Counts for 60 second				Corrected counts	ln(I_t)
t	t '= ρt	I	II	III	Mean(I_0)	$I_t=I_0-I_b$	
0	0	20695	20962	20658	20771.6	20756	9.94
0.019	0.149	683	614	614	637	621.34	6.43
0.038	0.299	25	25	31	27	11.34	2.42
0.057	0.448	30	20	23	24.33	8.67	2.15
0.076	0.598	21	18	20	19.66	4	1.38

1.Graph of I_t v/s t' ;

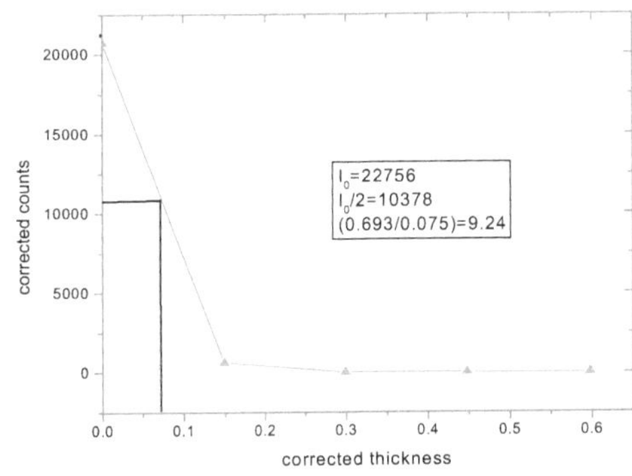

2.Graph ln (I_t) v/s corrected thickness (t');

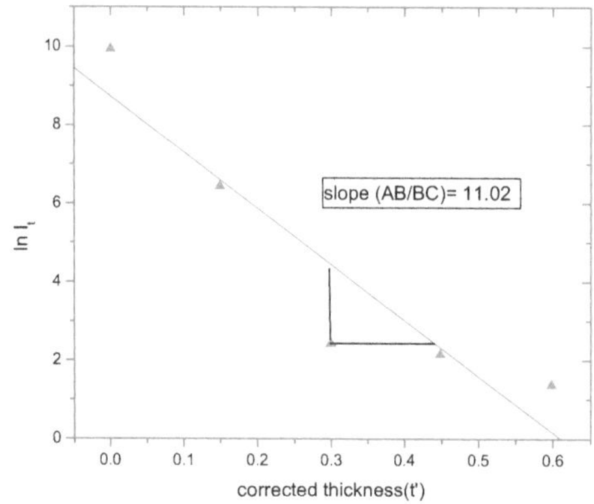

Mass Attenuation coefficient ;

μ_m = 9.24 cm^2/gm   (From graph 1)

μ_m = 11.02 cm^2/gm   (From graph 2)

Linear attenuation;

$\mu = \rho \times \mu_m$

  = 7.87 x 9.24

μ = 72.71 cm^{-1}

iii) Attenuation of beta particle from Tl-204 using Steel as target;

Back Ground Counts for 60 seconds:

I	II	III
23	17	25

Mean	21.66

Density of Steel = 7.8 gm/cm^3.

Thickness In cm.	Corrected Thickness in gm/cm^2	Counts for 60 second				Corrected counts	$\ln(I_t)$
t	t '= ρt	I	II	III	Mean(I_0)	$I_t=I_0-I_b$	
0	0	22368	22261	22276	22301.66	22280	10.01
0.012	0.936	2253	2218	2228	2233	2211.34	7.70
0.028	0.2184	59	53	52	54.66	33	3.49
0.043	0.3354	25	30	23	26	4.34	1.467
0.055	0.429	24	22	32	26	4.34	1.467

1. Graph of I_t v/s t' ;

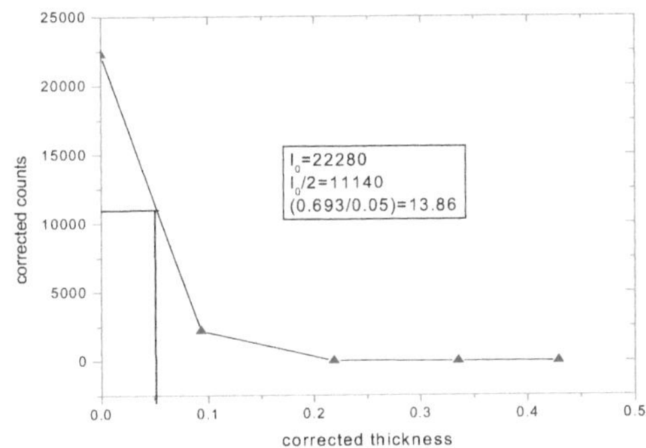

2.Graph of $\ln(I_t)$ v/s corrected thickness (t');

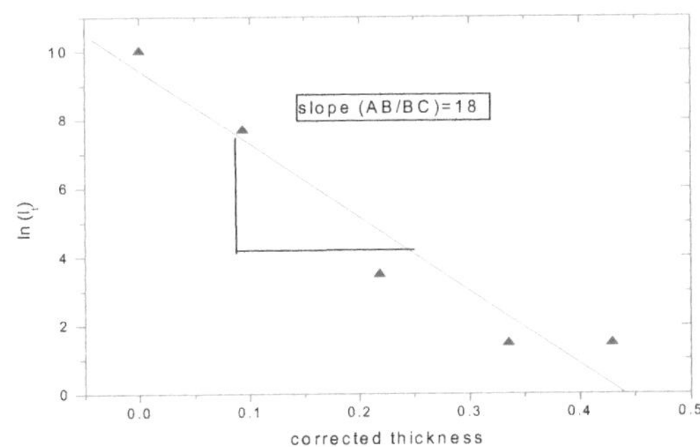

Mass Attenuation coefficient ;

μ_m = 13.86 cm^2/gm   (From graph 1)

μ_m = 18 cm^2/gm   (From graph 2)

Linear attenuation;

$\mu = \rho \times \mu_m$

   = 7.8 x 13.86

   = 108.10 cm^{-1}

iv) Attenuation of beta particle from Tl-204 using Brass as target;

Back Ground Counts for 60 seconds:

I	II	III
15	15	17

Mean	15.66

Density of Brass(Cu$_3$Zn$_2$) = 8.470 gm/cm^3.

Thickness In cm.	Corrected Thickness in gm/cm^2	Counts for 60 second				Corrected counts	ln(I$_t$)
t	t ' = ρt	I	II	III	Mean(I$_0$)	I$_t$=I$_0$-I$_b$	
0	0	23130	23007	23166	23101	23085.34	10.04
0.005	0.042	10650	10730	10631	10670.33	10654.67	9.273
0.01	0.084	3923	3883	3890	3898.33	3882.67	8.264
0.015	0.124	1057	1084	1064	1068.33	1052.67	6.959
0.02	0.169	306	304	314	308	292.34	5.677
0.025	0.211	79	81	76	78.66	63	4.143

1.Graph of I_t v/s t' ;

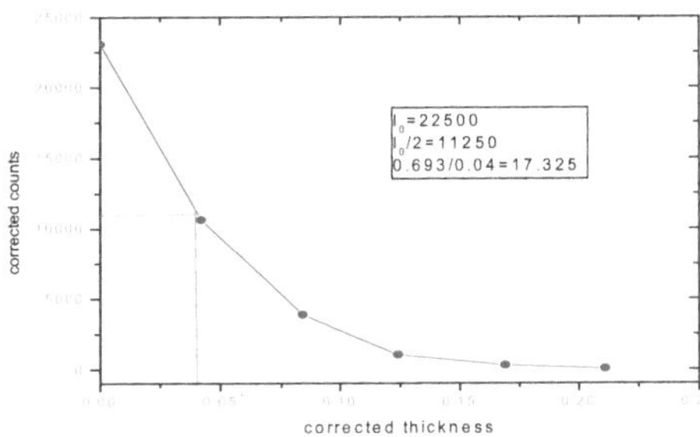

2.Graph of ln I_t v/s corrected thickness (t');

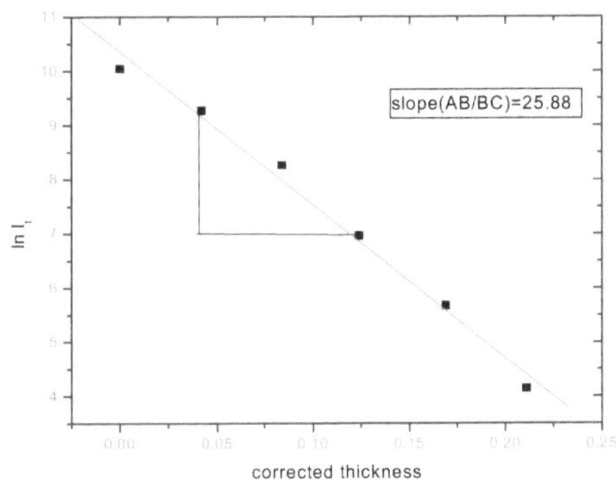

Mass Attenuation coefficient ;

μ_m =   17.325 cm^2/gm (From graph 1)

μ_m =   25.88 cm^2/gm (From graph 2)

linear   attenuation;

$\mu = \rho \times \mu_m$

$= 8.470 \times 17.325$

$= 146.74 \text{ cm}^{-1}$

v) Attenuation of beta particle from Tl-204 using Copper as target;

Back Ground Counts for 60 seconds:

I	II	III
19	16	21

Mean	18.66

Density of Cu = 8.96 gm/cm^3.

Thickness In cm.	Corrected Thickness in gm/cm^2	Counts for 60 seconds				Corrected counts	ln(I$_t$)
t	t '= ρt	I	II	III	Mean(I$_0$)	I$_t$=I$_0$-I$_b$	
0	0	22904	22868	22863	22878.33	22859.67	10.037
0.0025	0.0224	15676	15835	15860	15790.33	15771.67	9.665
0.005	0.044	9581	9480	9434	9498.33	9479.67	9.1569
0.0075	0.067	5367	5415	5294	5445.33	5426.67	8.599
0.01	0.089	2808	2862	2996	2888.66	2870	7.962
0.012	0.107	1487	1487	1472	1482	1463.34	7.288

1.Graph of I_t v/s t' ;

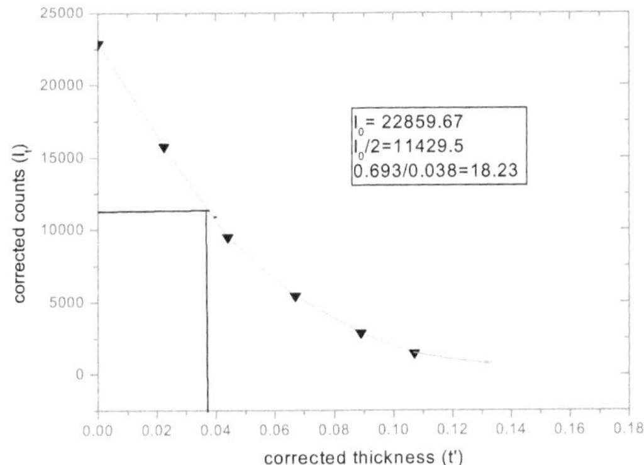

2.Graph of ln It v/s corrected thickness (t');

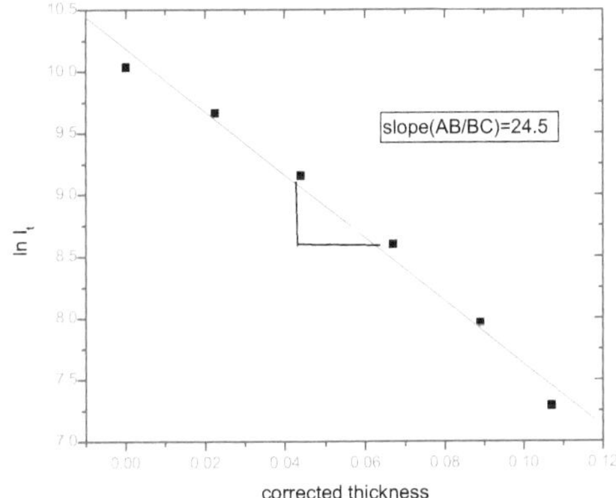

Mass Attenuation coefficient ;

μ_m = 18.23 cm^2/gm   (From graph 1)

μ_m = 24.5   cm^2/gm   (From graph 2)

linear attenuation ;

$\mu = \rho \times \mu_m$

   = 8.96 x 18.23

   = 163.34 cm^{-1}

vi) Attenuation of beta particle from Tl-204 using Aluminium as target;

Back Ground Counts for 60 seconds:

I	II	III
23	17	25

Mean	21.66

Density of Al = 2.7 gm/cm^3.

Thickness In cm.	Corrected Thickness in gm/cm^2	Counts for 60 seconds				Corrected counts	ln(I_t)
t	t '= ρt	I	II	III	Mean(I_0)	$I_t = I_0 - I_b$	
0	0	22898	22607	22674	22726	22704.34	10.03
0.02	0.054	7089	6907	6929	6975	6953.34	8.84
0.04	0.108	1207	1189	1204	1200	1178.34	7.07
0.06	0.162	139	128	117	128	106.34	4.66
0.08	0.216	31	26	41	32.66	11	2.397
0.10	0.270	32	28	34	31.33	9.67	2.26
0.12	0.324	24	38	30	30.66	9	2.19

1.Graph of I_t v/s t' ;

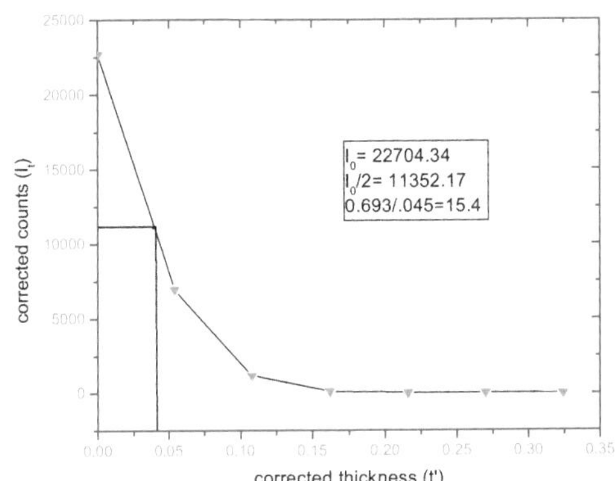

2.Graph of ln I_t v/s corrected thickness (t');

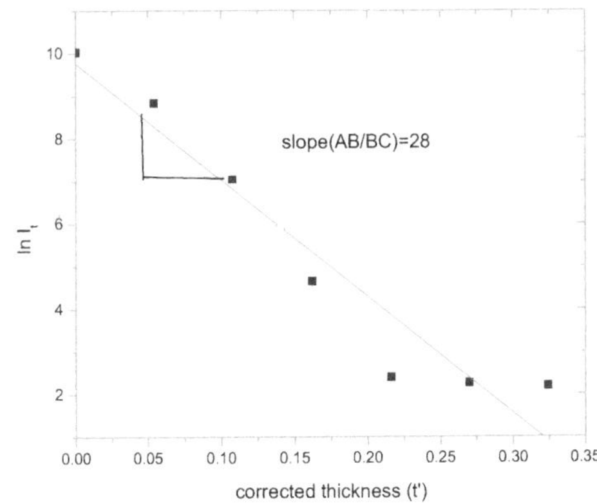

Mass Attenuation coefficient ;

μ_m = 15.4 cm^2/gm (From graph 1)

μ_m =   28 cm^2/gm (From graph 2)

Linear attenuation ;

$$\mu = \rho \times \mu_m$$

$$= 2.7 \times 15.4$$

$$= 41.58 \ cm^{-1}$$

COMPARISON OF MASS ATTENUATION CO-EFFICIENT WITH DENSITY OF MATERIALS ;

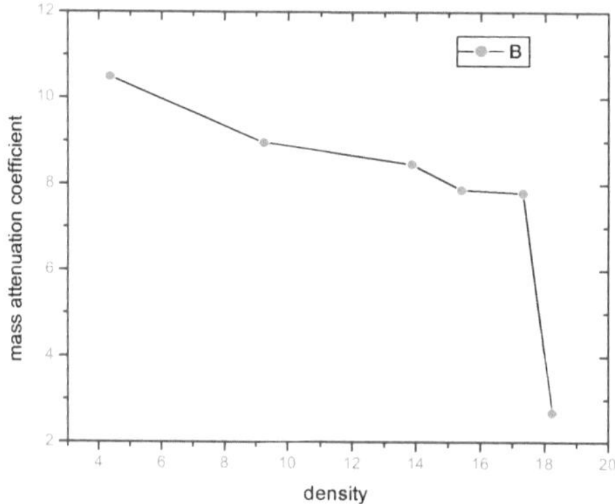

The above graph shows that the mass-attenuation depends on the density of the given materials . It can be noted from the graph that, as the density (ρ) of the given materials increases there is decrease in the mass-attenuation coefficient (μ_m) of the corresponding materials for the attenuation of the given (β-radiations) radioactive radiations.

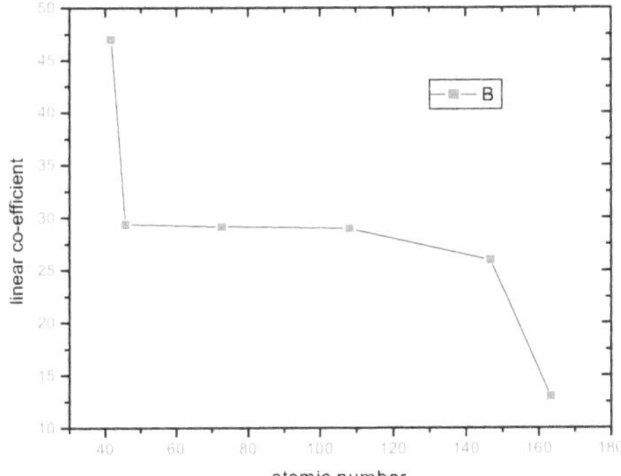

From the above graph it is clear that the linear attenuation co-efficient of the given materials for the absorption of beta-radiations decreases with the increase in the atomic number(Z) of the corresponding materials.

5.1.4: Attenuation of Beta particles in compounds:

Compounds used	Effective atomic number(Z)	Density of compound In gm/cm^3
Al_2SO4	27.64	2.672
$BaSO_4$	45.4	4.50
$MgSO_4$	11.86	2.66
NaCl	16.308	2.17
$ZnSO_4$	29.07	3.54

i) Attenuation of beta particle from Tl-204 using Al_2SO_4 as target;

Back Ground Counts for 60 seconds:

I	II	III
14	18	15

Mean	15.66

Density of Al_2SO_4 = 2.672 gm/cm^3.

Thickness In cm.	Corrected Thickness in gm/cm^2	Counts for 60 seconds				Corrected counts	ln(I_t)
t	t '= ρt	I	II	III	Mean(I_0)	I_t=I_0-I_b	
0	0	22726	22791	22590	22702.23	22686.57	10.02
0.023	0.061	1573	1547	1630	1583.33	1567.67	7.35
0.046	0.122	247	224	282	251	235.34	5.46
0.069	0.184	38	31	40	36.33	20.67	3.02
0.092	0.245	37	20	27	28	12.34	2.51
0.115	0.307	23	19	23	21.66	6	1.79

1.Graph of I_t v/s t' ;

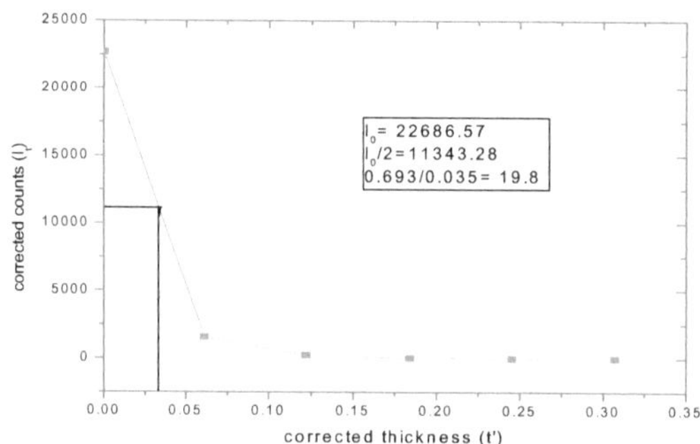

2.Graph of ln I_t v/s corrected thickness (t');

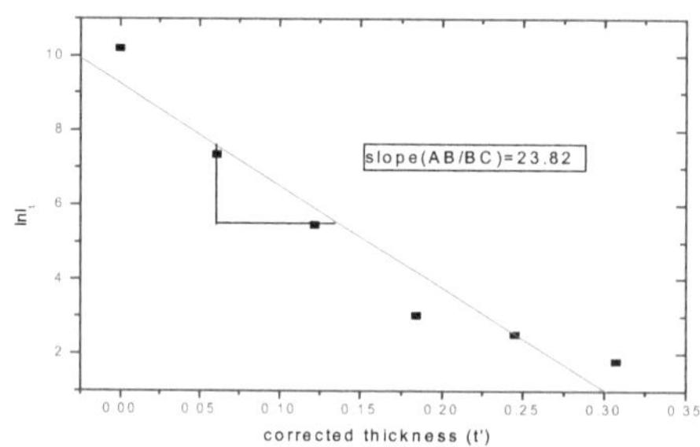

Mass Attenuation coefficient ;

$\mu_m =$ 19.8 cm^2/gm   (From graph 1)

$\mu_m =$ 23.82 cm^2/gm   (From graph 2)

Linear attenuation ;

$\mu = \rho \times \mu_m$

$= 2.672 \times 19.8 = 52.90$ cm^{-1}

ii) Attenuation of beta particle from Tl-204 using BaSO$_4$ as target;

Back Ground Counts for 60 seconds:

I	II	III
11	20	16

Mean	15.66

Density of BaSO$_4$ = 8.470 gm/cm^3.

Thickness In cm.	Corrected Thickness in gm/cm^2	Counts for 60 seconds				Corrected counts	ln(I$_t$)
t	$t' = \rho t$	I	II	III	Mean(I$_0$)	I$_t$=I$_0$-I$_b$	
0	0	22484	22580	22853	22639	22623.36	10.026
0.038	0.171	737	752	780	756.33	740.67	6.607
0.043	0.193	33	41	34	36	20.34	3.012
0.0445	0.200	22	23	21	22	6.34	1.846
0.047	0.211	21	30	32	27.66	12	2.484
0.049	0.220	26	30	23	26.33	10.67	2.367

1.Graph of I_t v/s t' ;

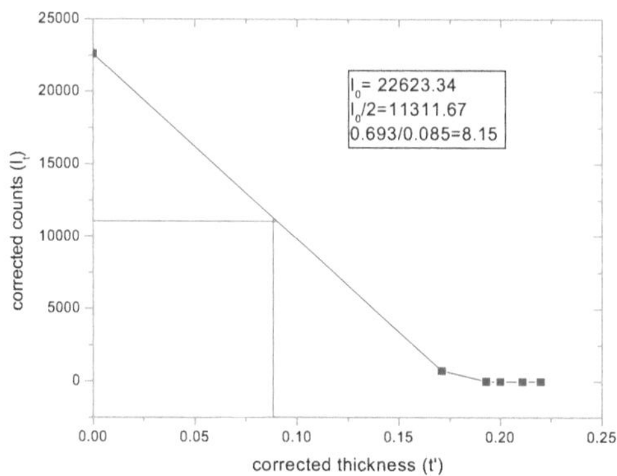

2.Graph of ln I_t v/s corrected thickness (t');

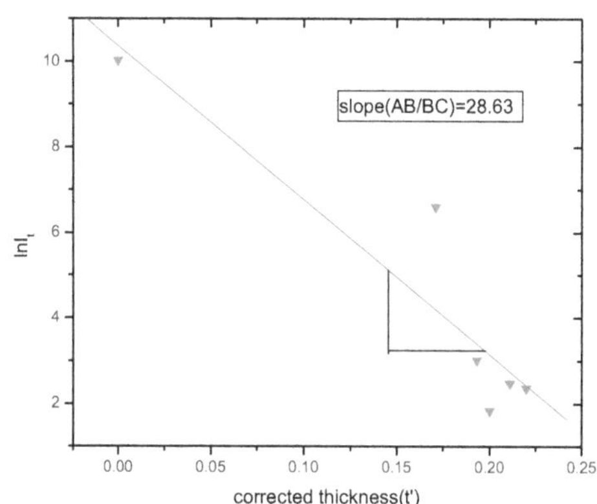

Mass Attenuation coefficient ;

μ_m = **18.15 cm^2/gm**   **(From graph 1)**

μ_m = **28.63 cm^2/gm**   **(From graph 2)**

Linear attenuation ;

$$\mu = \rho \times \mu_m$$

$$= 4.50 \times 18.15$$

$$= 81.675 \text{ cm}^{-1}$$

iii) Attenuation of beta particle from Tl-204 using $MgSO_4$ as target;

Back Ground Counts for 60 seconds:

I	II	III
15	15	17

Mean	15.66

Density of $MgSO_4$ = 2.66 gm/cm^3.

Thickness In cm.	Corrected Thickness in gm/cm^2	Counts for 60 seconds				Corrected counts	$\ln(I_t)$
t	t '= ρt	I	II	III	Mean(I_0)	$I_t = I_0 - I_b$	
0	0	23130	23007	23166	23101	23085.34	10.06
0.005	0.042	10650	10730	10631	10670.33	10654.67	9.23
0.01	0.084	3923	3883	3890	3898.33	3882.67	8.24
0.015	0.124	1057	1084	1064	1068.33	1052.67	6.959
0.02	0.169	306	304	314	308	292.34	5.677
0.025	0.211	79	81	76	78.66	63	4.143

1.Graph of I_t v/s t' ;

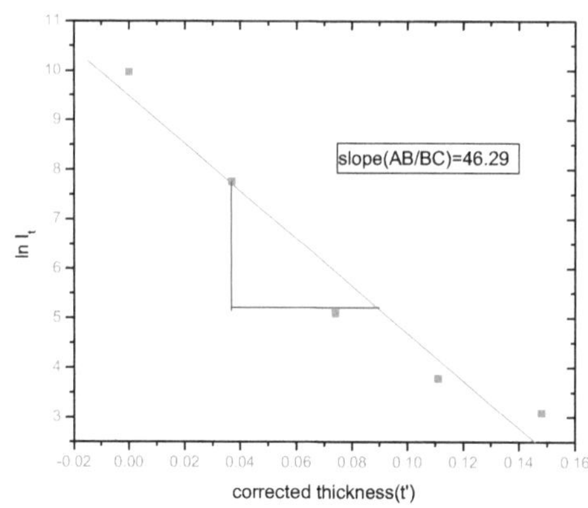

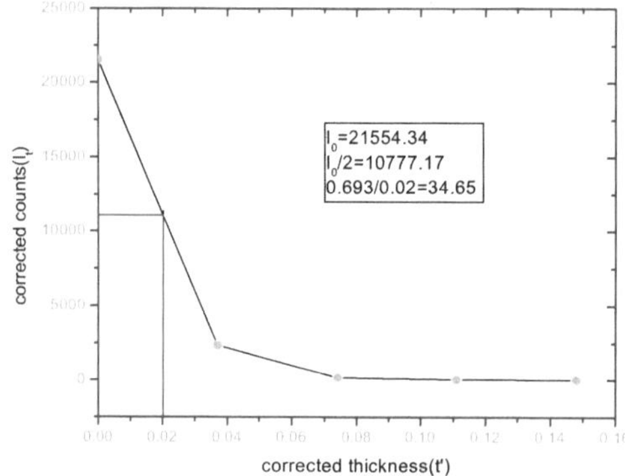

2.Graph of ln I_t v/s corrected thickness (t');

Mass Attenuation coefficient ;

μ_m = 34.65 cm^2/gm   (From graph 1)

μ_m = 46.29 cm^2/gm   (From graph 2)

Linear attenuation ;

$\mu = \rho \times \mu_m$

$\quad = 2.66 \times 34.65$

$\quad = 92.16 \text{ cm}^{-1}$

iv) Attenuation of beta particle from Tl-204 using NaCl as target;

Back Ground Counts for 60 seconds:

I	II	III
16	17	24

Mean	19

Density of NaCl = 2.17 gm/cm³.

Thickness In cm.	Corrected Thickness in gm/cm²	Counts for 60 seconds				Corrected counts	$\ln(I_t)$
t	t '= ρt	I	II	III	Mean(I_0)	I_t=I_0-I_b	
0	0	21105	21137	21102	21114.66	21095.66	9.95
0.04	0.86	819	803	860	827.33	808.33	6.69
0.08	0.173	41	72	66	59.66	40.66	3.70
0.12	0.260	26	30	24	26.66	7.66	2.03
0.16	0.347	30	19	26	25	6	1.79
0.20	0.434	20	25	24	23	4	1.38

1.Graph of I_t v/s t' ;

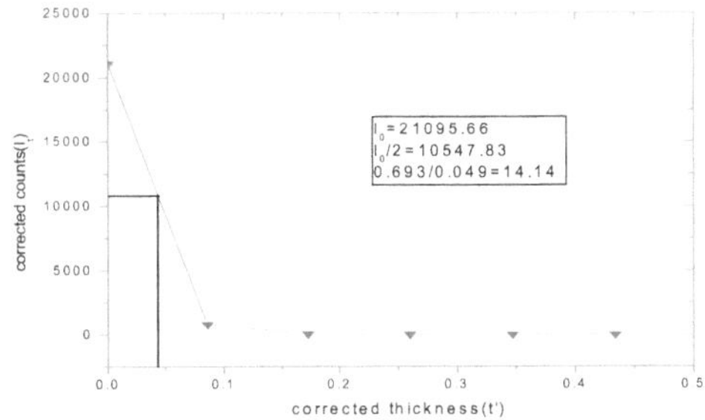

2.Graph of ln I_t v/s corrected thickness (t');

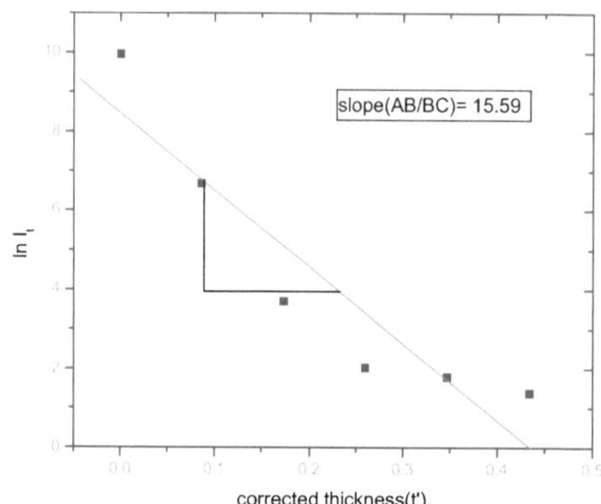

Mass Attenuation coefficient ;

μ_m = 14.14 cm^2/gm   (From graph 1)

μ_m = 15.59 cm^2/gm   (From graph 2)

Linear attenuation ;

$\mu = \rho \times \mu_m$

= 2.17 x 14.14

= 30.68 cm^{-1}

v) Attenuation of beta particle from Tl-204 using NaCl as target;

Back Ground Counts for 60 seconds:

I	II	III
20	22	15

Mean	19

Density of $ZnSO_4 = 3.54gm/cm^3$.

Thickness In cm.	Corrected Thickness in gm/cm^2	Counts for 60 seconds				Corrected Counts	$\ln(I_t)$
t	$t' = \rho t$	I	II	III	Mean(I_0)	$I_t = I_0 - I_b$	
0	0	12607	12421	12859	12629	12610	9.442
0.047	0.166	2097	2057	2081	2078.33	2059.33	7.630
0.094	0.332	101	114	97	104	85	4.44
0.188	0.665	29	32	29	30	11	2.3978
0.376	1.331	28	27	28	27.66	8.66	2.159
0.752	2.662	16	27	24	22.33	3.33	1.203

1.Graph of I_t v/s t' ;

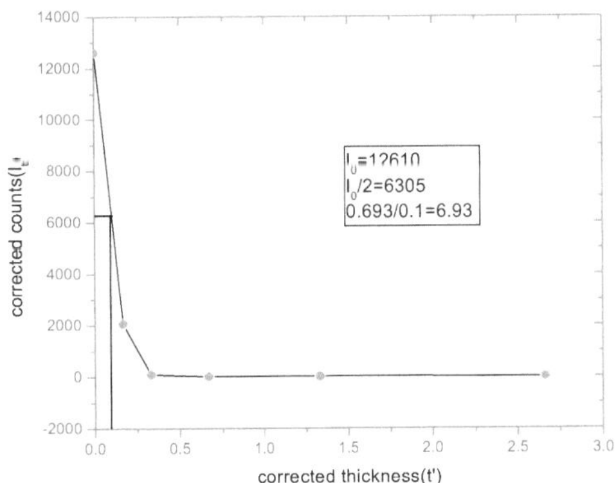

2.Graph of ln I_t v/s corrected thickness (t');

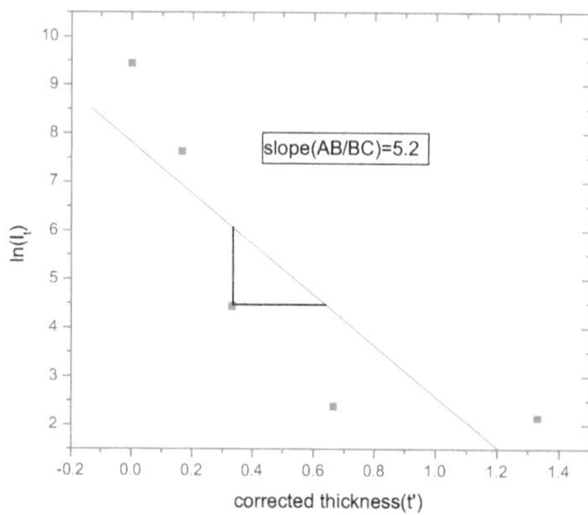

Mass Attenuation coefficient ;

μ_m = 6.93 cm^2/gm   (From graph 1)

μ_m = 5.2 cm^2/gm   (From graph 2)

Linear attenuation ;

$\mu = \rho \times \mu_m$

   = 3.54 x 6.93

   = 24.53 cm^{-1}

COMPARISON OF MASS ATTENUATION CO-EFFICIENT WITH DENSITY OF COMPOUNDS ;

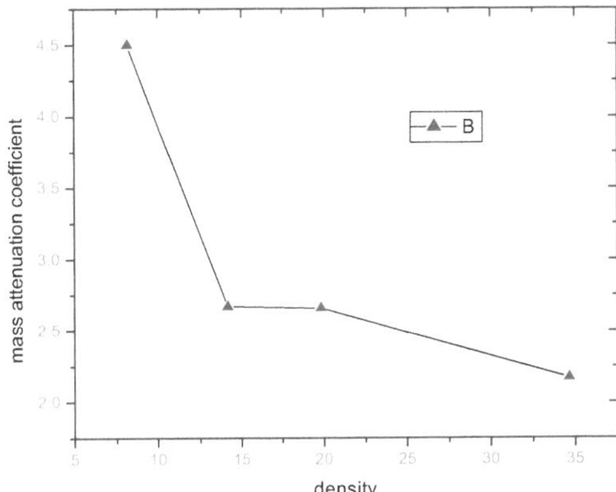

The above graph will shows that the mass attenuation coefficient   will change with density of the compounds that is,as the density (ρ) of compounds increases the mass attenuation c0-efficient(μ_m) will decrease. Here for a few values of density the mass attenuation is found to remain constant which is not true and this may be due to background or instability of the detector.

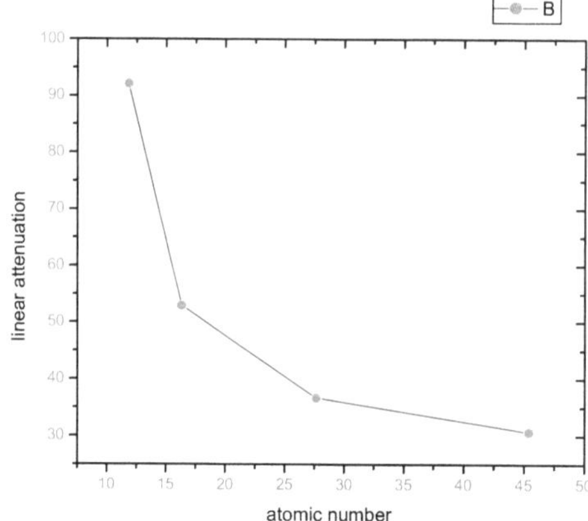

As discussed above in case of metals, the same dialogue follows for the compounds too,i.e., the linear attenuation co-efficient(μ) of the compounds depends on the effective atomic number(Z_{eff}) of the given compounds.

RESULT :

For Metals:

Metals	Mass attenuation co-efficient (μ_m)	Linear attenuation (μ)
Silver	4.35 cm^2 gm^{-1}	45.63 cm^{-1}
Iron	9.24 cm^2 gm^{-1}	72.71 cm^{-1}
Steel	13.86 cm^2 gm^{-1}	108.10 cm^{-1}
Brass	17.325 cm^2 gm^{-1}	146.74 cm^{-1}
Copper	18.23 cm^2 gm^{-1}	163.34 cm^{-1}
Aluminum	15.4 cm^2 gm^{-1}	41.58 cm^{-1}

For Compounds:

Compounds	Mass attenuation co-efficient (μ_m)	Linear attenuation (μ)
Al_2SO_4	19.8 $cm^2 gm^{-1}$	52.90 cm^{-1}
$BaSO_4$	18.15 $cm^2 gm^{-1}$	81.675 cm^{-1}
$MgSO_4$	34.65 $cm^2 gm^{-1}$	92.16 cm^{-1}
$NaCl$	14.14 $cm^2 gm^{-1}$	30.68 cm^{-1}
$ZnSO_4$	6.93 $cm^2 gm^{-1}$	24.53 cm^{-1}

Experiment 2:

A} AIM:To study the Bremsstrahlung radiations emitted due to interaction of beta-particles with matter.

Block Diagram of Experimental Arrangement:

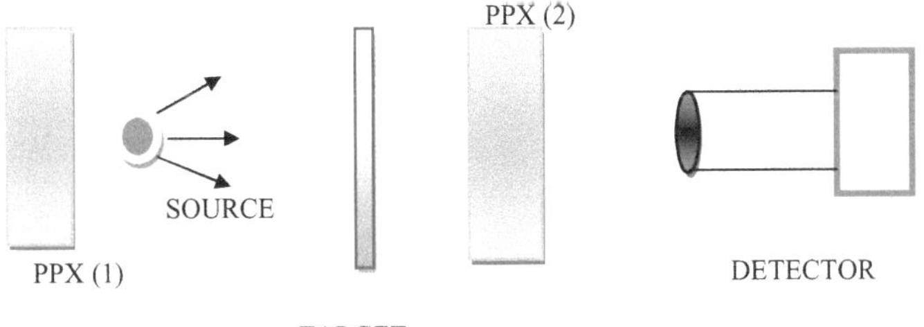

The above block diagram shows the experimental arrangement with target (metal or compound) in position 1. Similarly the target is kept after the Perspex (2) for the counts in position 2.

1) Bremsstrahlung radiations emitted from copper using beta source Tl-204:

Target	Thickness in mm		Counts for 180 seconds				EB counts $I=C_1-C_2$
			In position 1 (C_1)		In position 2 (C_2)		
	0.0224	0.0025	112926	111976	111208	111050	101322
Copper	0.005	0.044	108393	108028	103977	103978	4233
	0.0075	0.067	106787	106328	104744	104819	1776
	0.01	0.09	102668	102012	101778	101955	473.5

Here I is the number of external bremsstralung radiations emitted from copper and the counts (C_2) is the internal bremsstrahlung radiations emitted due to the interaction of beta rays within the source material.

B} Aim:

- To find the Z dependence of external bremsstrahlung using different materials.
- To find the Yield constant (K), Material constant(Σ_B) and Index factor(n) for different materials using different Sources.

Appartus:

- Scintillation counter.
- Targets of different thickness.
- Perspex

Formulae:

$$I=KZ^n (t/A) \, e^{-\Sigma t}$$

Where   K=yield constant

        Σ_B=material constant

        n=index factor

        Z=atomic number

        A=atomic mass

Experimental set up

Diagram:

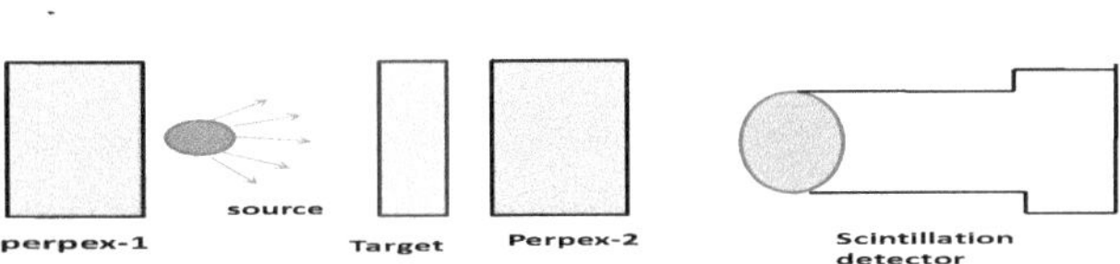

Materials Used:

1) **Perspex 1:** This is the non conducting material, which is used to stop the radiations leaking in to the air.

2) Source: It is the radioactive and purely a beta emitting material.

Sources used: ^{204}Tl & Sr^{90}-Y^{90}

Energy of source:763.76keV &

3)Materials used:

Metals/compounds	Atomic number
Aluminium	13
Copper	29
Silver	47
Brass	29.55
Steel	29.18
Iron	26

4) Perspex 2: In bremsstrahlung process when high energy electrons are incident on the high Z material the electrons are slowed down due to high positive charge of nucleus and the electron is slowed down or is completely absorbed in the target material. Electrons escaped from the target are absorbed by this perspex.

Procedure:

Depending upon the radiations to attenuated, we place perspex in two different positions.

Position 1:

This is the position where the perspex is placed between target and the detector. When the beta particles are incident on the target external bremsstrahlung effect is produced. By placing the perspex in position-1 the electrons due to the external bremstrahlung are attenuated. At this position counts obtained are noted as C_1.

Position 2:

This is the position where perspex is placed between the source and target. As we study that bremsstrahlung is due to the high energy electrons interact with the high Z atom. Bremsstrahlung is produced within the source which is called as internal bremsstrahlung. By placing perspex in position 2 we attenuate the electrons produced in bremsstrahlung. The counts obtained are noted as C_2.

2. The difference between the two counts is taken i.e $C_1 - C_2$ which is the actual EB count (I).

3. Repeat this procedure for different materials of different thicknesses.

4. Plot the graph of external bremsstrahlung counts versus thickness of target. The slope obtained by this graph is noted as Σ_B

5. From the grph of external bremsstrahlung counts versus thickness the intercept at y-axis is noted as lan(KZn).

6. Repeat this procedure for different targets of different thicknesses to get $lan(kz^n)$.

7. Plot te graph of $lan(kz^n)$. Obtained for different materials verses the lnZ values of the corresponding materials(or compounds), which is straight line whose slope gives the index factor 'n'.

Observations:

I)   Using Tl-204 as source for Z-dependence of external bremsstrahlung :

Target:

1: Aluminium(Al)

Target thickness in mm	Target thickness in mg/cm^2	Counts for 180 seconds						Actual counts I=C1-C2	ln(IA/t)
		Counts for position 1 : C1			Counts for position 2 : C2				
		I	II	mean	I	II	Mean		
0.05	13.5	18228	18903	18565	18142	18289	18215	350	5.25
0.10	27	18055	18048	18050	18433	18051	18242	191	4.8
0.15	40.5	17706	18069	17887	18050	17844	17947	60	3.68
0.20	54	17715	17845	17780	17979	17638	17808	37	2.9

Graph:

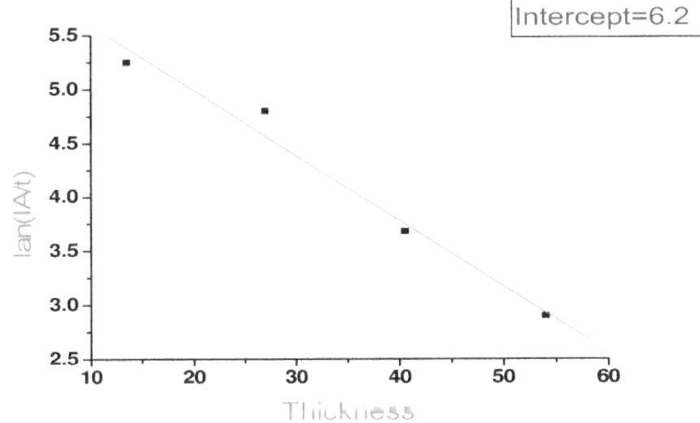

From graph: $lanKZ^n = 6.2$

$\Sigma_B = -0.06052$

2. Iron(Fe)

Target thickness in mm	Target thickness in mg/cm²	Counts for 180 seconds						Actual counts I=C1-C2	ln(IA/t)
		Counts for position 1 : C1			Counts for position 2 : C2				
		I	II	mean	I	II	Mean		
0.17	133.79	36872	36872	36872	36531	36531	36531	341	6.54
0.34	267.7	32591	32591	32591	32511	36531	36531	80	5.6
0.51	401.37	30209	30209	30209	29625	29625	29625	584	4.29
0.68	535.16	27303	27303	27303	26737	26739	26739	566	4.07

Graph:

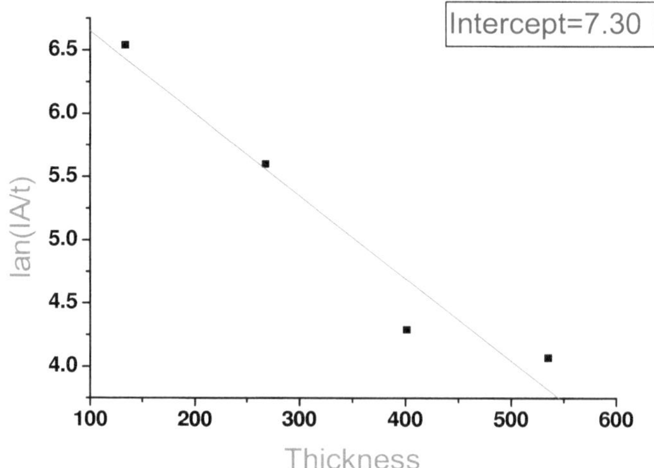

From graph: $\text{lan} KZ^n = 7.3$

$\Sigma_B = -0.00652$

3: Copper(Cu)

Target thickness in mm	Target thickness in mg/cm^2	Counts for 180 seconds						Actual counts $I=C1-C2$	$\ln(IA/t)$
		Counts for position 1 : C1			Counts for position 2 : C2				
		I	II	Mean	I	II	Mean		
0.04	35.68	39757	39757	39757	39248	39248	39248	509	6.68
0.06	53.52	37621	37621	37621	37181	37181	37181	440	6.21
0.08	71.36	37089	38066	37577	36488	36524	36506	547	6.15
0.10	89.2	36453	36453	36453	35775	35775	35775	566	6.17
0.12	107.0	34575	34575	34575	34344	34344	34344	231	4.92

Graph:

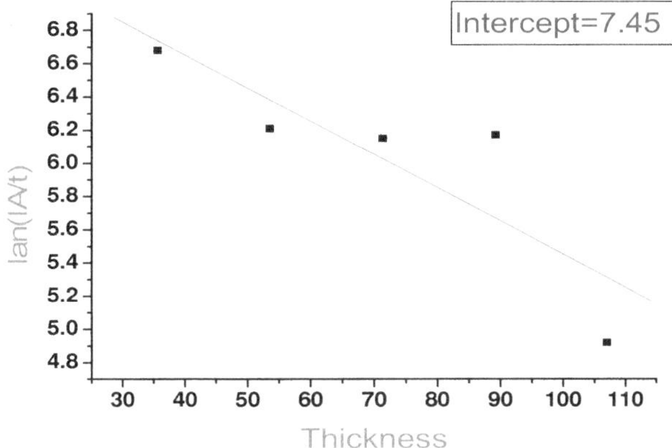

From graph: $\text{lan}KZ^n = 7.45$

$\Sigma_B = -0.01996$

4:Brass(Cu_3Zn_2)

Target thickness in mm	Target thickness in mg/cm^2	Counts for 180 seconds						Actual counts I=C1-C2	ln(IA/t)
		Counts for position 1 : C1			Counts for position 2 : C2				
		I	II	Mean	I	II	Mean		
0.04	34	115253	116692	115972	115906	114783	115344	628	7.68
0.08	68	107143	107944	107543	106610	105610	106110	1434	6.43
0.12	102	101400	101753	101576	98702	98456	98579	2997	6.76
0.16	136	96589	96191	96390	91974	92255	92114	4276	6.83

Graph:

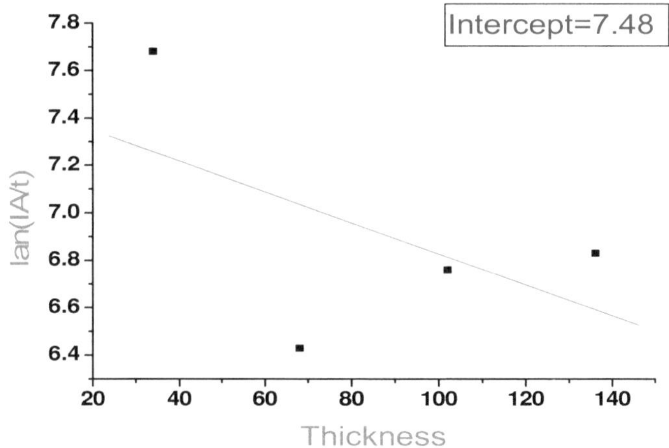

From graph: $lanKZ^n = 7.48$

$\Sigma_B = -0.00653$

5:Steel(FeZn$_3$)

Target thickness in mm	Target thickness in mg/cm^2	Counts for 180 seconds						Actual counts I=C1-C2	ln(IA/t)
		Counts for position 1 : C1			Counts for position 2 : C2				
		I	II	Mean	I	II	Mean		
0.13	100.75	17005	16554	16779	16800	16426	16613	166	6.82
0.26	201.5	15705	15720	15713	15593	15358	15476	237	5.5
0.39	302.25	15524	14271	14898	14675	14324	14499	399	5.63
0.52	403	13385	13758	13572	13770	13607	13688	117	4.32

Graph:

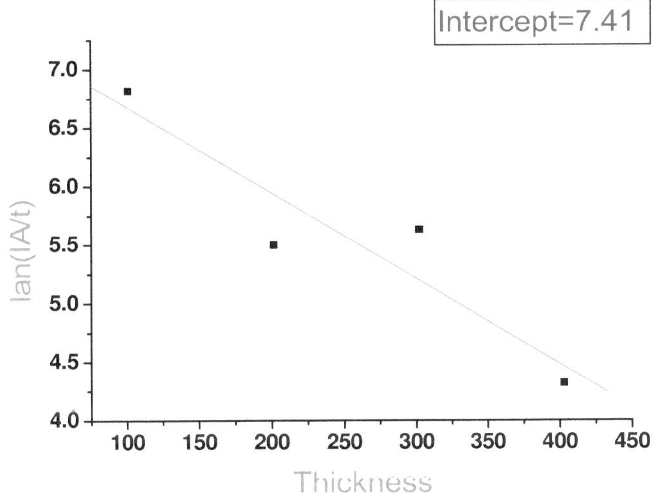

From graph: $lanKZ^n = 7.41$

$\Sigma_B = -0.007323$

6:Silver(Ag)

Target thickness in mm	Target thickness in mg/cm^2	Counts for 180 seconds						Actual counts I=C1-C2	ln(IA/t)
		Counts for position 1 : C1			Counts for position 2 : C2				
		I	II	Mean	I	II	Mean		
0.03	31.47	43694	43694	43694	42998	42998	42998	696	7.87
0.06	62.94	42961	42961	42961	42231	42231	42231	730	7.13
0.09	94.4	42772	42772	42772	41942	41942	41942	830	6.85

Graph:

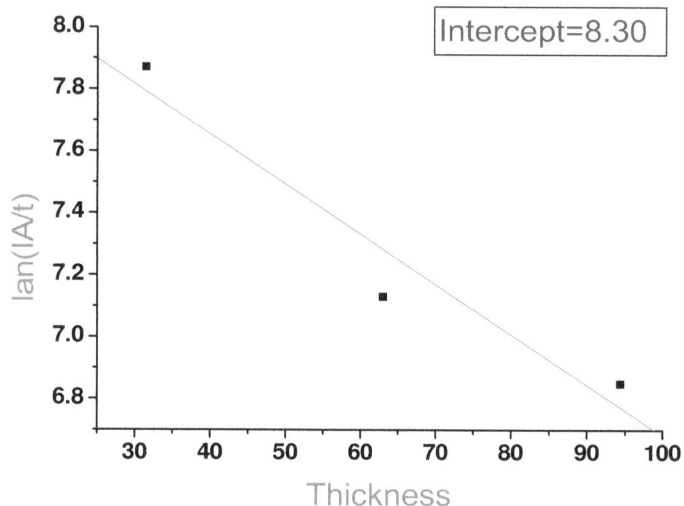

From graph: $lanKZ^n$ =8.3

Σ_B=-0.01621

Table to study about material constant and yield constant

Sl no.	Target	atomic number (Z)	lanZ	(Intercept) lanKZ^n	Material constant (Σ_B)	Yield contant (K)
1	Aluminium	13	2.56	6.2	-0.006052	5.99
2	Iron	26	3.25	7.3	-0.00652	5.46
3	Copper	29	3.36	7.45	-0.01996	5.26
4	Steel	29.18	3.37	7.41	-0.007323	5.02
5	Brass	29.55	3.38	7.48	-0.00653	5.01
6	Silver	47	3.85	8.3	-0.01621	5.00

Graph: lanZ varies with lanKZ^n

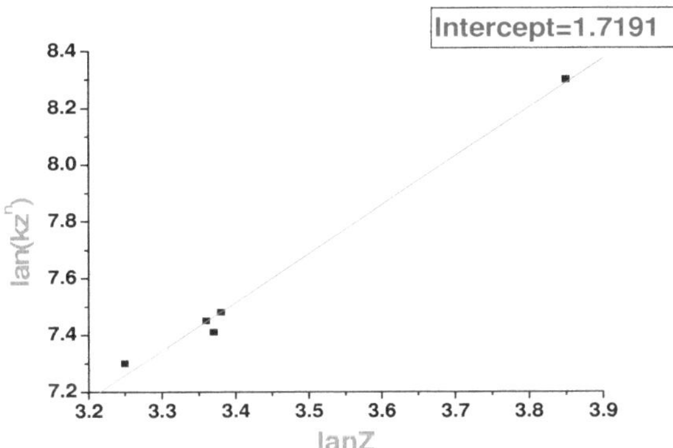

The slope of the above graph gives the Index factor(n) = 1.7191

Graph: Yield constant K varies with atomic number Z

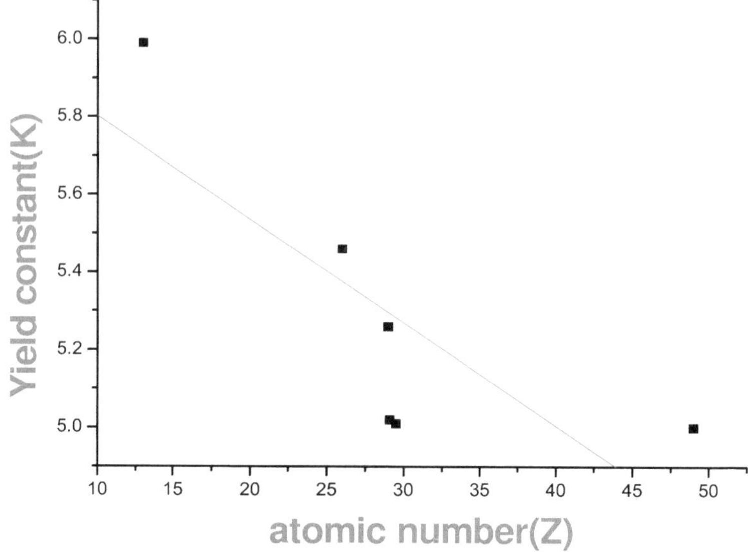

Using Strontium($Sr_{90} - Y_{90}$) as source

1. Lead(Pb-82)

Target thickness in mg/cm^2	Counts for 180 seconds						Actual counts I=C1-C2	ln(IA/t)
	Counts for position 1 : C1			Counts for position 2 : C2				
	I	II	Mean	I	II	Mean		
116.74	5701	5701	5701	5191	5191	5191	510	6.80
233.48	6377	6377	6377	5971	5971	5971	406	5.86
350.2	6528	6528	6528	5955	5955	5955	573	5.82
466.76	6429	6429	6429	5981	5981	5981	448	5.29
583.7	6474	6474	6474	6020	6020	6020	454	5.08
700.44	6531	6531	6531	6146	6146	6146	385	4.73

Graph:

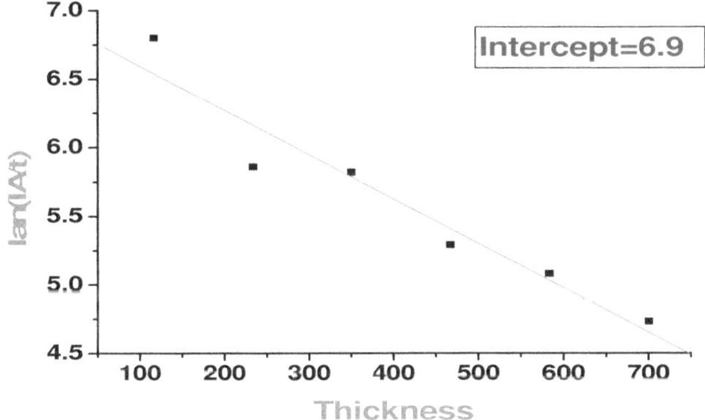

From graph: $lanKZ^n$ = 6.91

Σ_B= -0.00324

2. Nickel (Ni-28)

Target thickness in mg/cm^2	Counts for 180 seconds						Actual counts I=C1-C2	ln(IA/t)
	Counts for position 1 : C1			Counts for position 2 : C2				
	I	II	Mean	I	II	Mean		
129.2	6373	6373	6373	6129	6129	6129	244	4.70
258.4	6500	6500	6500	6178	6178	6178	322	4.29
387.6	6332	6332	6332	6151	6151	6151	181	3.31
516.8	6019	6019	6019	5772	5772	5772	247	3.33
646	6165	6165	6165	6003	6003	6003	162	2.68

Graph:

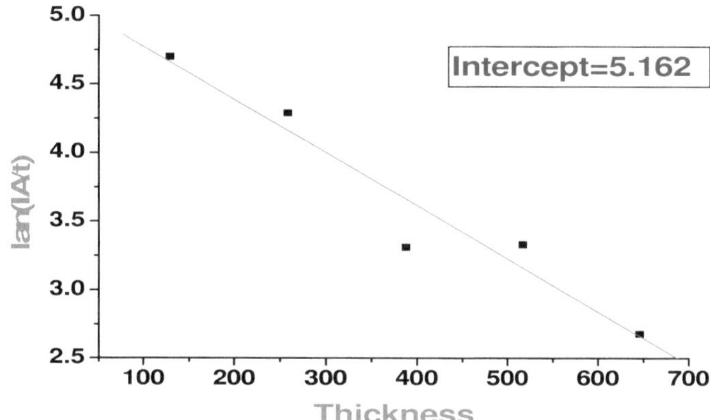

From graph: $lanKZ^n$ = 5.162

Σ_B = -0.00387

3.Copper(Cu-29)

Target thickness in mg/cm^2	Counts for 180 seconds						Actual counts I=C1-C2	ln(IA/t)
	Counts for position 1 : C1			Counts for position 2 : C2				
	I	II	Mean	I	II	Mean		
42.2	3827	3827	3827	3772	3772	3772	55	4.41
144.58	3778	3778	3778	3657	3657	3657	121	3.97
223.24	3843	3843	3843	3663	3663	3663	180	3.93
295.35	3798	3798	3798	3586	3586	3586	212	3.83
367.72	3778	3778	3778	3542	3542	3542	236	3.70
474.33	3847	3847	3847	3593	3593	3593	254	3.50
549.138	3849	3849	3849	3582	3582	3582	267	3.43

Graph:

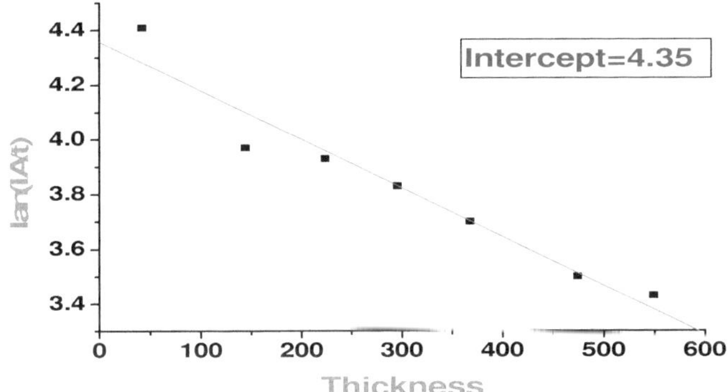

From graph: $lanKZ^n = 4.35$

$\Sigma_B = -0.00178$

4.Silver(Ag-47)

Target thickness in mg/cm^2	Counts for 180 seconds						Actual counts I=C1-C2	ln(IA/t)
	Counts for position 1 : C1			Counts for position 2 : C2				
	I	II	Mean	I	II	Mean		
116	3600	3600	3600	3528	3600	3600	72	4.20
232	3636	3636	3636	3522	3636	3636	114	3.97
348	3664	3664	3664	3499	3664	3664	165	3.93
464	3779	3779	3779	3579	3779	3779	200	3.83
580	3645	3645	3645	3430	3645	3645	315	4.07
696	3629	3629	3629	3430	3629	3629	199	3.42

Graph:

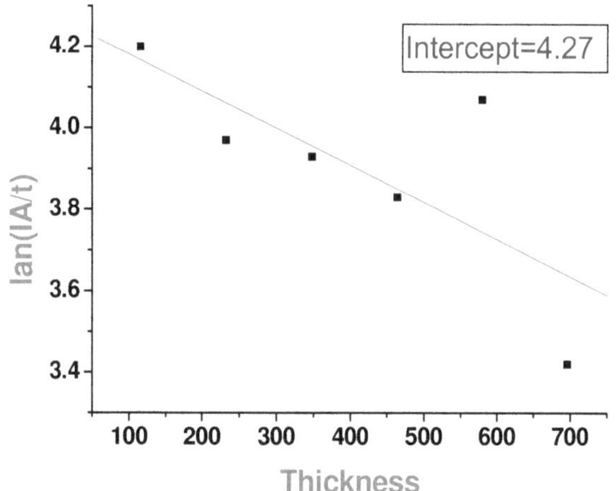

From graph: $\text{lan}KZ^n = 4.27$

$\Sigma_B = -0.00091133$

Table to study about material constant and yield constant

Sl no.	Target	atomic number (Z)	lanZ	(Intercept) $lankz^n$	Material constant (Σ_B)	Yield constant (K)
1	Nickel	28	3.33	5.162	-0.00387	0.43
2	Copper	29	3.36	4.35	-0.00178	0.18
3	Silver	47	3.85	4.27	-0.000911	0.076
4	Lead	82	4.40	6.91	-0.00324	0.37

Graph: lanZ varies with lanKZ^n

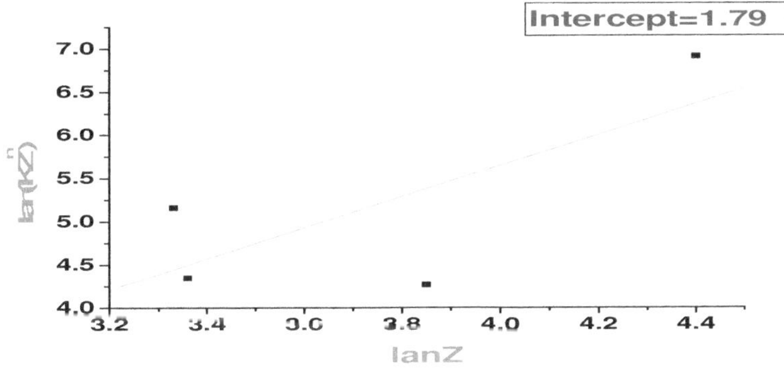

The slope of the above graph gives the Index factor(n)=1.79

Graph: Yield constant K varies with atomic number Z

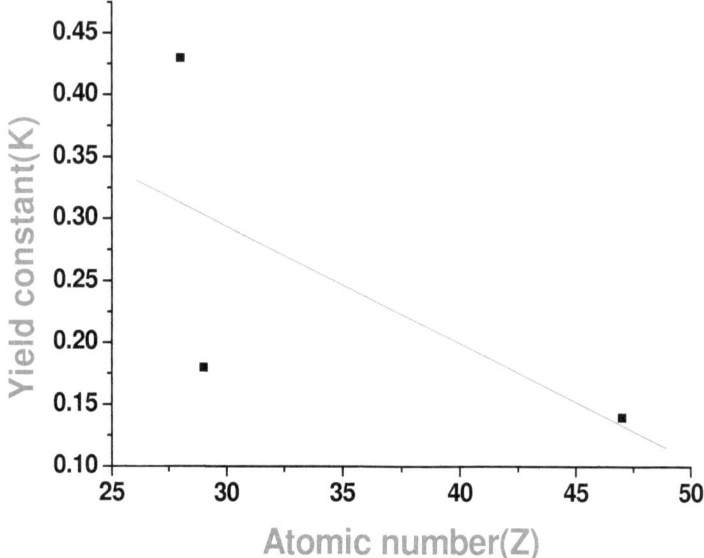

RESULT

1. **The Z-dependence of different material has been studied using source Tl_{204} and Sr_{90}-Y_{90}**

2. **Material constant(k), Yield constant($\sum_B$) and Index factor(n) are calculated for different materials and different sources.**

3. **The obtained value of Index factor(n) from experiment is equal to 1.79, which is in agreement with standard value 1.9.**

CONCLUSION:

It is well known that any charged particle or a neutral particles when enters into target material, loses its energy either by ionizing exciting the atom or the particle may be completely absorbed by the material atoms.

In the present study, we have used the beta source Tl-204. These beta particles when incident on the different materials (metals or compounds) interact with the atoms of the target materials by the three processes as stated above. The beta particles are incident on the target material and depending on the decrease in the counts of the particles detected for increasing thickness of the target, the mass attenuation coefficient for the absorbing of beta particles for different materials have been found. The mass attenuation for different metals as well as for different compounds and compounds salts have been studied. By knowing the mass attenuation coefficient the linear attenuation coefficient of the corresponding metal or compound has been found. Also the dependence of mass attenuation coefficient of metals and compounds on the density of the respective metals and compounds have been successfully studied.

We also know that when an electron is incident on a target (metal or compound), it interacts with the coulombic field of the nucleus which deaccelerates the electron or beta particle giving rise to "bremsstrahlung radiations". The bremsstrahlung radiations may also arise due to interaction of beta particles within the source atoms. The external bremsstrahlung radiations emitted due to the interaction of beta particles with the atoms of the target have been detected and studied.

BIBLIOGRAPHY

1. Atomic & Nuclear physics : S.N. Ghoshal – S.Chand & Company (1998)
2. Nuclear Physics: D.C.Tayal – Himalaya Publishing House (2009)
3. Introduction to Nuclear physics: Herald .A. Enge
4. Nuclear physics : R.D.Evans
5. Bremsstrahlung Radiations of Beta particles : Paper by C.S.Mahajan
6. External Bremsstrahlung Radiation : Paper by T.K.Umesh , M.V.Manjunath, L.Vinay Kumar

YOUR KNOWLEDGE HAS VALUE

- We will publish your bachelor's and
  master's thesis, essays and papers

- Your own eBook and book -
  sold worldwide in all relevant shops

- Earn money with each sale

Upload your text at www.GRIN.com
and publish for free